AF379427

ANIMALS OF THE MOUNTAINS
Alpine Ibex
by Kaitlyn Duling
BLASTOFF! READERS
2
BELLWETHER MEDIA • MINNEAPOLIS, MN

Blastoff! Readers are carefully developed by literacy experts to build reading stamina and move students toward fluency by combining standards-based content with developmentally appropriate text.

Level 1 provides the most support through repetition of high-frequency words, light text, predictable sentence patterns, and strong visual support.

Level 2 offers early readers a bit more challenge through varied sentences, increased text load, and text-supportive special features.

Level 3 advances early-fluent readers toward fluency through increased text load, less reliance on photos, advancing concepts, longer sentences, and more complex special features.

★ **Blastoff! Universe**

Reading Level

Grade
K

Grades
1–3

Grade
4

This edition first published in 2021 by Bellwether Media, Inc.

No part of this publication may be reproduced in whole or in part without written permission of the publisher. For information regarding permission, write to Bellwether Media, Inc., Attention: Permissions Department, 6012 Blue Circle Drive, Minnetonka, MN 55343.

Library of Congress Cataloging-in-Publication Data

Names: Duling, Kaitlyn, author.
Title: Alpine Ibex / Kaitlyn Duling.
Description: Minneapolis, MN : Bellwether Media, 2021. | Series: Blastoff! readers : Animals of the mountains | Includes bibliographical references and index. | Audience: Ages 5-8 | Audience: Grades K-1 | Summary: "Relevant images match informative text in this introduction to alpine ibex. Intended for students in kindergarten through third grade"-- Provided by publisher.
Identifiers: LCCN 2020041124 (print) | LCCN 2020041125 (ebook) | ISBN 9781644874134 (library binding) | ISBN 9781648340901 (ebook)
Subjects: LCSH: Ibex--Juvenile literature.
Classification: LCC QL737.U53 D848 2021 (print) | LCC QL737.U53 (ebook) |DDC 599.64/8--dc23
LC record available at https://lccn.loc.gov/2020041124
LC ebook record available at https://lccn.loc.gov/2020041125

Editor: Kieran Downs   Designer: Brittany McIntosh

Printed in the United States of America, North Mankato, MN.

Table of Contents

Life in the Mountains

Alpine ibex are wild goats. They live in the Alps of central Europe.

They roam between
the mountain forests
and snowy peaks.

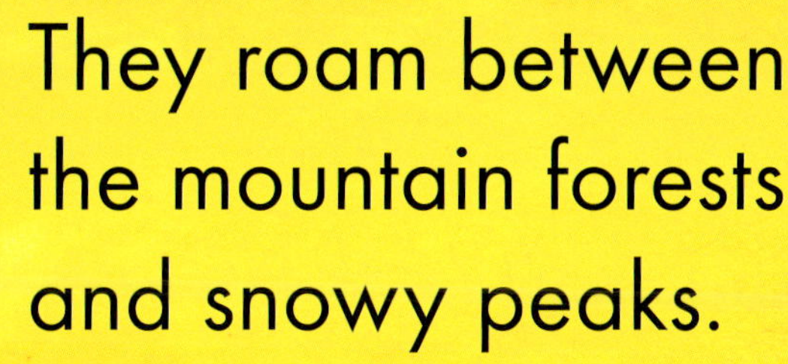

5

Alpine ibex have **adapted** to their **biome**. They use their **cloven hooves** to climb on steep rocks.

Their hooves have sharp edges that help them **balance**.

Alpine ibex have strong legs. They can jump more than 6 feet (2 meters) with no running start!

This helps them move around the steep mountains.

The mountains have many **predators**. But alpine ibex scare them away.

Special Adaptations

They show off their large, curved horns. They stand on their back legs.

Winters in the Alps are very cold. Alpine ibex grow thick, warm **coats** to stay warm.

When the weather warms, they **molt**. Some of the hair falls out.

Alpine Ibex Stats
Least Concern
Near Threatened
Vulnerable
Endangered
Critically Endangered
Extinct in the Wild
Extinct
conservation status: least concern
life span: up to 18 years
molting
13

Alpine ibex **migrate** twice each year. In winter, they move to south-facing mountainsides.

Snow melts faster in these areas.
Ibex can find food more easily.

As the snow melts in spring,
alpine ibex climb up the mountains.

They rest on rocky cliffs during the day. They **graze** as the sun goes down.

Finding Food

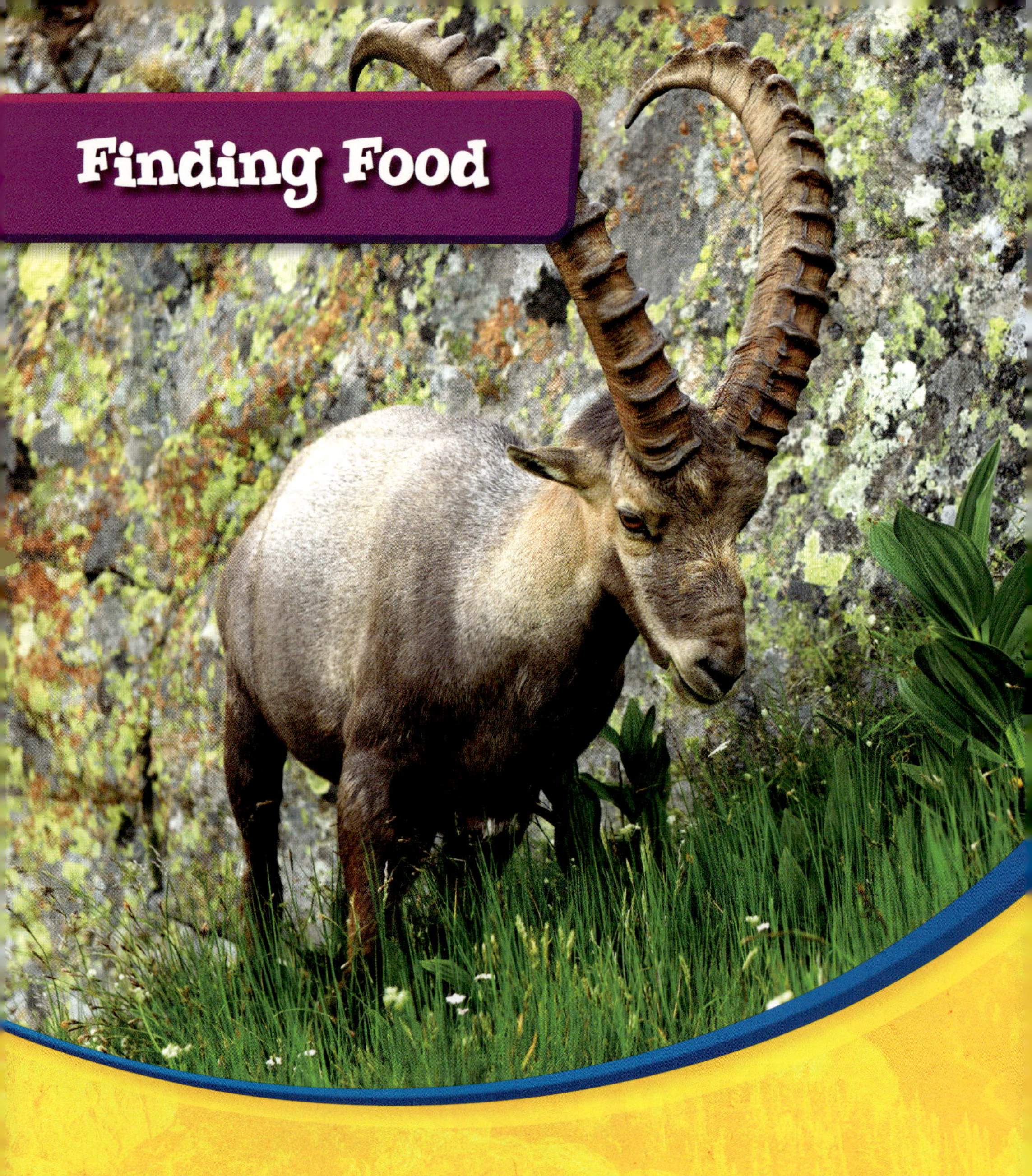

Alpine ibex are **herbivores**. They search for grasses, mosses, and leaves.

They stand on their hind legs
to reach leaves on trees.

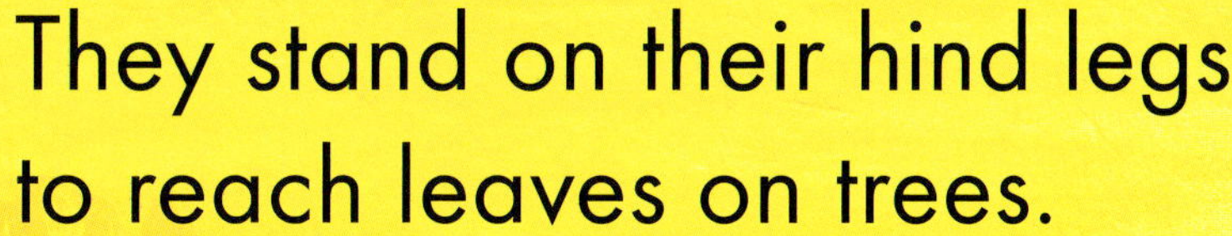

Alpine Ibex Diet

lady's slippers

English
oak leaves

alpine meadow grass

Alpine ibex need salt
to survive. They lick salt
found on mountain rocks.

These goats live well in the mountains!

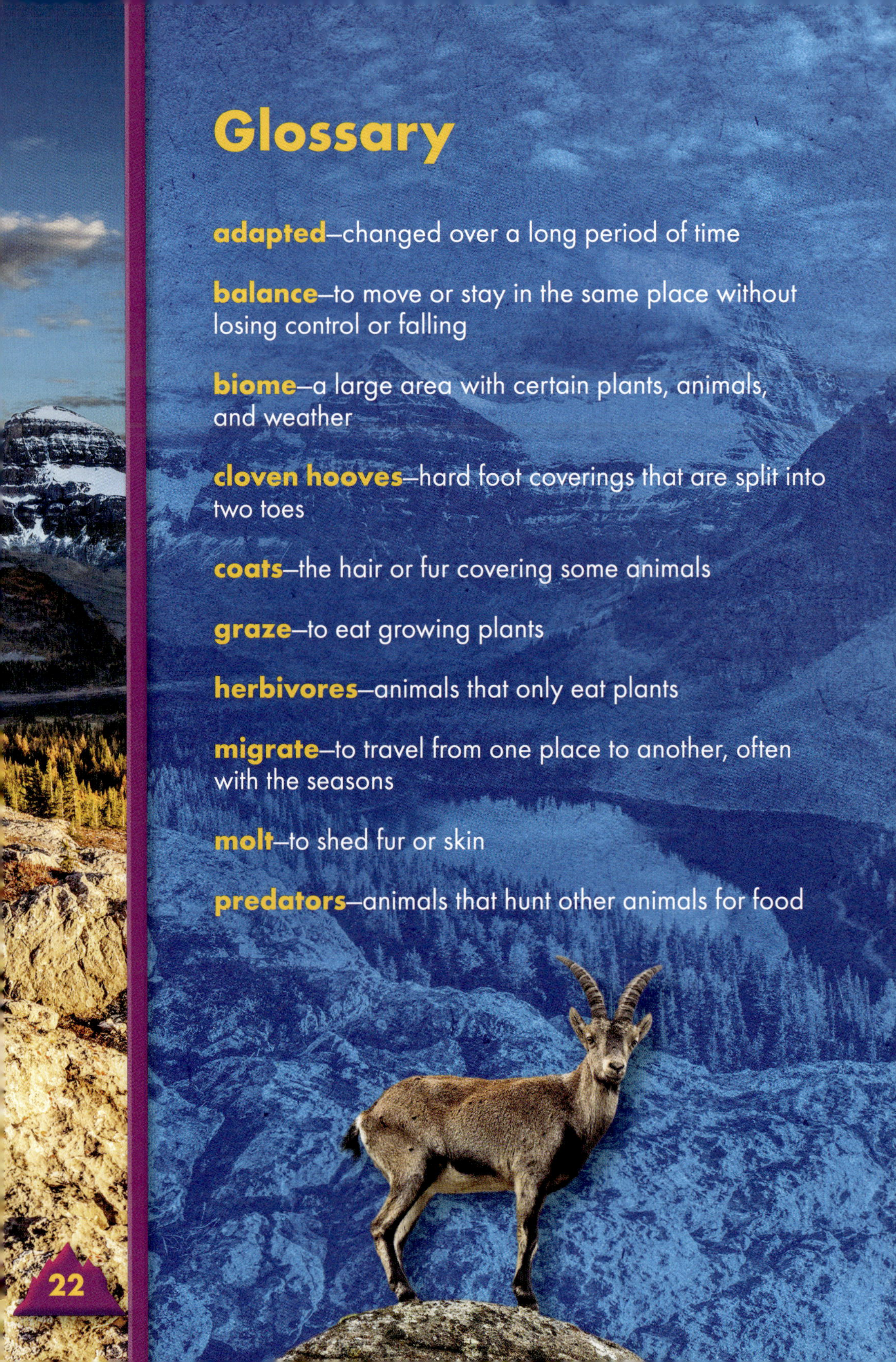

Glossary

adapted—changed over a long period of time

balance—to move or stay in the same place without losing control or falling

biome—a large area with certain plants, animals, and weather

cloven hooves—hard foot coverings that are split into two toes

coats—the hair or fur covering some animals

graze—to eat growing plants

herbivores—animals that only eat plants

migrate—to travel from one place to another, often with the seasons

molt—to shed fur or skin

predators—animals that hunt other animals for food

To Learn More

AT THE LIBRARY

Davies, Monika. *How High Up the Mountain?: Mountain Animal Habitats.* Mankato, Minn.: Amicus Illustrated, 2019.

Pettiford, Rebecca. *Mountains.* Minneapolis, Minn.: Jump!, 2018.

Shaffer, Lindsay. *Mountain Goats.* Minneapolis, Minn.: Bellwether Media, 2020.

ON THE WEB

Factsurfer.com gives you a safe, fun way to find more information.

1. Go to www.factsurfer.com.

2. Enter "alpine ibex" into the search box and click 🔍.

3. Select your book cover to see a list of related content.

Index